My First Book of Zoo Animals

Mike Unwin

Illustrated by

Daniel Howarth

A & C BLACK

AN IMPRINT OF BLOOMSBURY

LONDON NEW DELHI NEW YORK SYDNEY

Published 2014 by A & C Black, an imprint of
Bloomsbury Publishing Plc, 50 Bedford Square
London, WC1B 3DP

www.bloomsbury.com

ISBN: 978-1-4729-0531-4

A CIP catalogue for this book is available from the British Library.

This book is produced using paper that is made from wood grown in managed, sustainable forests. It is natural, renewable and recyclable. The logging and manufacturing processes conform to the environmental regulations of the country of origin.

Printed and bound in China by Leo Paper Products.

Modern zoos look after
animals carefully and try
to give them everything
they need. Keeping anim
in captivity helps scientis
to understand how they

Contents

Looking aft

Zoos are exciting plac
to visit. But they also
important job. They h
to learn more about a
and to protect ones th
become rare.

White rhinos nearly be
Zoos gave them a saf

Giraffe

Aha! It's a giraffe. And that snake is really its tongue.

A giraffe reaches high into the treetops with its long neck and plucks juicy leaves with its long tongue.

Giraffes live on the plains of Africa. Their height means that they are quick to spot danger, such as lions.

Who is it?

What a lovely thick scarf. It's keeping somebody's nose nice and warm.

But who do you think it belongs to?

Scimitar-horned oryx

It's a scimitar-horned oryx.

Those horns look just like special Arabic swords, called scimitars. That's how this rare antelope got its name.

It comes from the Sahara Desert in Africa, where its pale colours help it to blend in. But today it is found only in zoos.

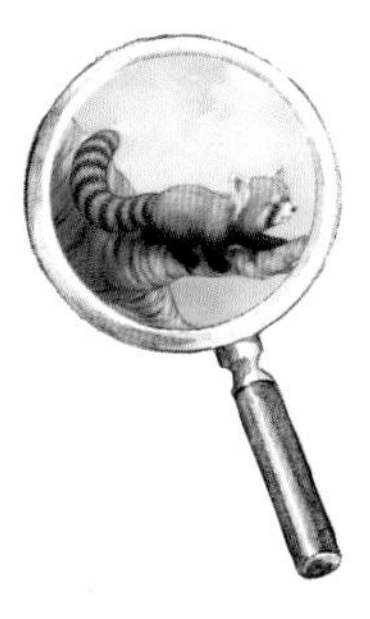

Who is it?

Look up in that tree.

Something red and furry is curled up on a branch.

But what is it?

Red panda

It's a red panda.
Now you can see its black paws and stripy face.

Red pandas are much smaller than giant pandas.
But like giant pandas, they live in China
and eat bamboo.

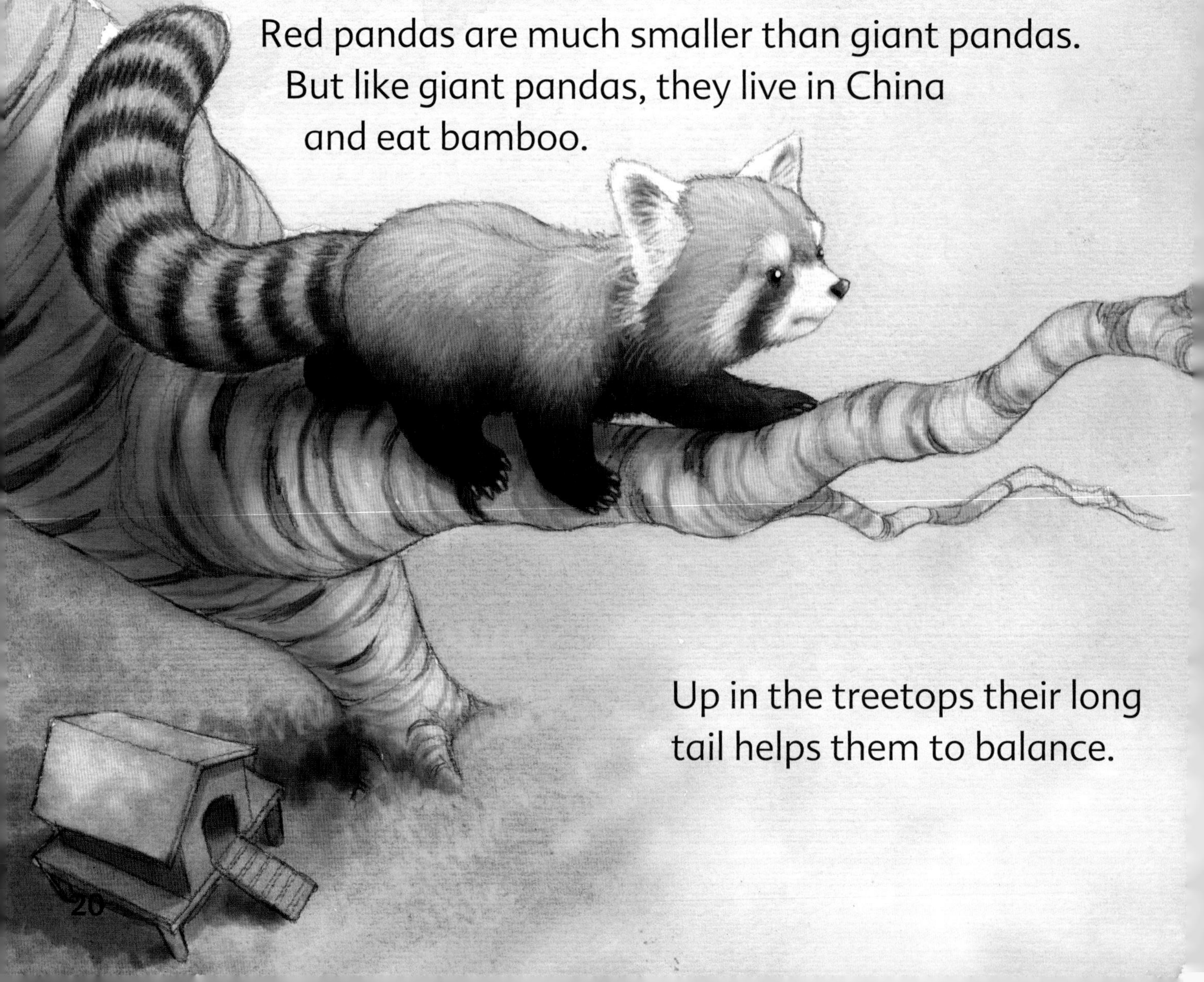

Up in the treetops their long
tail helps them to balance.

Who is it?

Whose heads are those?

Three small animals are standing up
to get a better view.

But they've turned their backs on you.

Meerkat

Aha! One has turned round to take a look at you.

Do you recognise that face? It's a meerkat, of course!

Meerkats live together in the deserts of South Africa. They help each other out by taking turns to look for danger. Standing on their back legs gives them a better view.

Who is it?

Something is playing with that old tyre. Something black and hairy, with long arms.

It must be very strong to lift up that tyre all by itself.

What do you think it is?

Gorilla

It's a gorilla. Gorillas live in small groups in the rainforests of Africa. They can weigh twice as much as a man.

Even though gorillas are very strong, they are very gentle and eat only plants.
Baby gorillas
love to play.

In the wild gorillas are
very rare. Zoos help
to protect them.

Who is it?

Look at those long stripy things waving in the breeze.

What do think they are?

They look a bit like flags, or scarves.

Ring-tailed lemur

Aha! They are tails. And they belong to ring-tailed lemurs.

Lemurs are related to monkeys but live only on the island of Madagascar.

Ring-tailed lemurs often climb in the trees.

When they come down to the ground they wave their tails like flags. This helps them to follow one another.

Who is it?

Look out! A strange animal is coming your way.

It has huge curved claws, like pirates' hooks.
And a long nose, with an extra-long tongue.

What can it be?

Giant anteater

It's a giant anteater. This South American animal uses its big claws to dig for ants. Then it sticks its long nose inside and licks them up with its sticky tongue. Delicious!

Can you see the baby on its back? Young giant anteaters ride on their mothers until they are ten months old.

Who is it?

Look at those claws clinging on tight.

Something has a tight grip on that branch. It must be a very good climber.

But what is it?

Koala

It's a koala sitting high up a tree.
Those claws help it cling on tight.

Koalas come from Australia,
where they eat the leaves of gum trees.

They are marsupials, just like kangaroos.
A mother keeps her baby in a special pouch.

Who is it?

Look out! Something has its eyes on you.
And it looks very fierce.

Are those its tufty ears sticking up?
Perhaps it's listening to you too.

What can it be?

Eagle owl

Aha! It's an eagle owl, the biggest owl in the world. Look at those sharp talons and beak for catching its prey.

Those tufts on its head are not really ears. They are feathers to help with camouflage.

During daytime the owl hides away. It is very hard to spot.

Who is it?

Two creatures are swimming
around underwater in their tank.

They look as fast and wriggly as fish.
But they've got beaks and feet.

What do you think they are?

Rockhopper penguin

They are rockhopper penguins.
Look at their fancy yellow head feathers.

On the islands where they live, these penguins hop about from one rock to another. That's how they get their name, of course!

In water, they are brilliant swimmers.

Who is it?

Look at that enormous, colourful beak.

What a weight it must be to carry about!

Who do you think it belongs to?

Toucan

It's a toucan.

This tropical bird lives in the jungles of South America. It uses its huge beak to pluck fruit from the tips of branches. It even reaches into tree holes to steal eggs.

The beak is not as heavy as it looks. A special honeycomb structure inside makes it very strong and light.

Who is it?

Something is coiled up behind those rocks.

It's long and shiny. And look, is that a flickering tongue?

What can it be?

Python

It's a python. It tastes the air with its forked tongue to help it find food.

Pythons are the longest snakes of all. They are not venomous. Instead they catch their prey by wrapping their coils around it. Then they swallow it whole.

Some pythons can grow to more than six metres. That's longer than three people laid end to end.

Who is it?

What a scary face!

This animal has got horns like a dinosaur. And it's climbing down the branch right towards you.

What can it be?

Chameleon

It's a chameleon. It may look like a monster but really it's perfectly harmless. Those horns are just for camouflage.

Chameleons move very slowly, gripping the branches with their toes and tail.

They can swivel their eyes and shoot out their tongue to catch insects. They can even change colour.

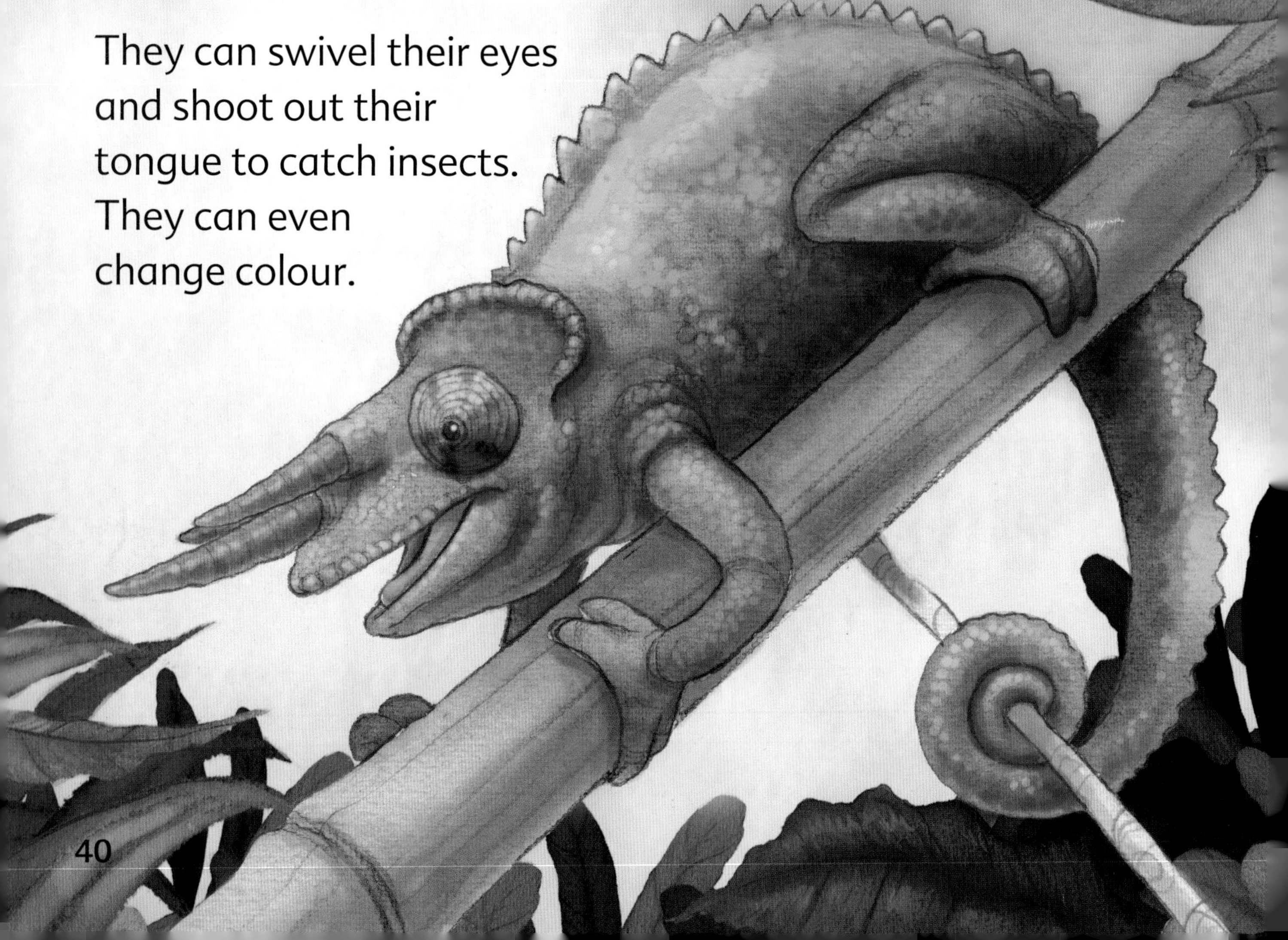

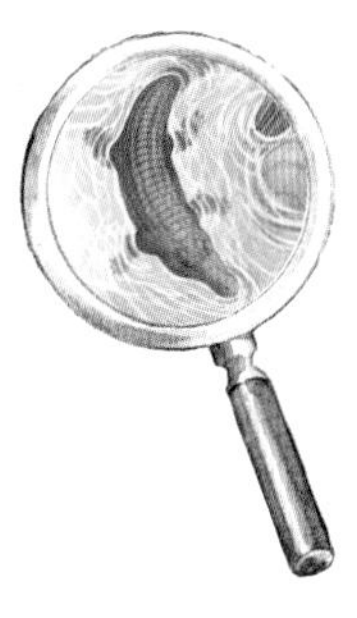

Who is it?

Something's floating in the water. And it's staring right at you.

Look at those bright yellow eyes and sharp teeth.

What do you think it can be?

Crocodile

It's a crocodile.
Now you can see its long tail and scaly back.

Crocodiles live in rivers.

They are very good swimmers, using their strong tail to push them through the water. Those sharp teeth help them to catch fish and other animals.

On sunny days crocodiles come onto land to warm up.

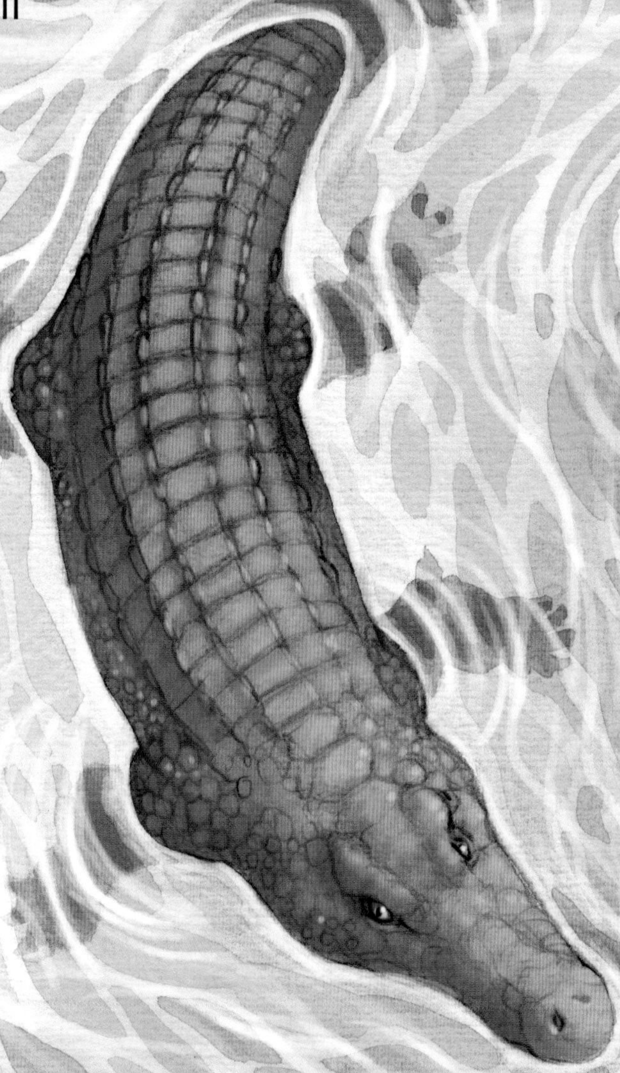

Who is it?

Is that tree moving all by itself?

Its leaves seem to be walking along the branch.

What's going on?

Leafcutter ants

Aha! It's a line of leafcutter ants.

These busy insects live in tropical rainforests. They cut off small pieces of leaf and carry them down to their underground home.

Under the ground the ants chew the pieces into mush. Soon fungus grows on it. This makes food for their babies.

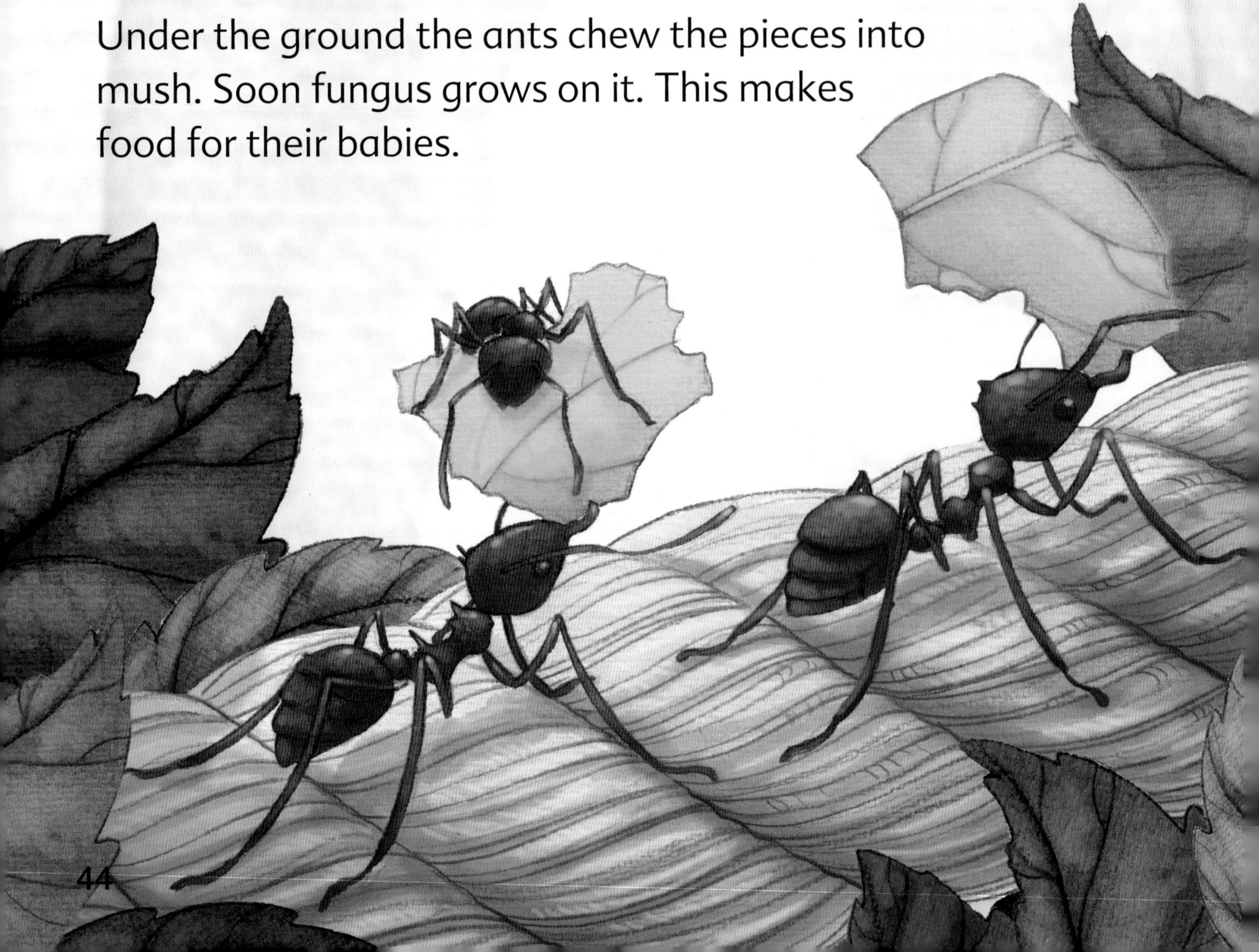

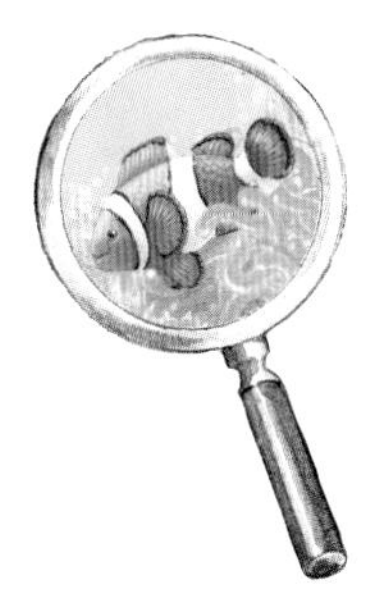

Who is it?

Something's hiding underwater.

Can you see its face peeking out from inside that sea anemone?

What do you think it is?

Clownfish

It's a clownfish. Now it's come out so you can take a good look.

These pretty little fish hide inside the stinging tentacles of sea anemones. But the anemones don't sting them. This keeps the clownfish safe from larger fish.

In return the clownfish eat up any small harmful creatures that might bother the anemone.

Zoo animal words

bamboo a tropical plant with tall, hollow stems that grows as tall as a tree.

camouflage patterns or shapes that help an animal blend into its background.

extinct gone forever, like the dinosaurs.

in captivity in an enclosure; not in the wild.

marsupial an animal, such as a kangaroo, that keeps its baby in a pouch.

rainforest thick forest in hot parts of the world where it rains all year round.

Visit the zoo

There are over 40 zoos and safari parks in Great Britain. Perhaps there is one near you. Here are three:

Bristol Zoo Gardens **London Zoo** **Edinburgh Zoo**

For details on all British zoos, visit *zoos-uk.com*.

Index